The Next Plague and How Science Will Stop It

WRITTEN BY

Alex Berezow, Ph.D.
Hank Campbell
Julianna LeMieux, Ph.D
Steve Schow, Ph.D.

A publication of the

AMERICAN COUNCIL
ON SCIENCE AND HEALTH

The Next Plague And What Science Will Do To Stop It

For more information, contact:

American Council on Science and Health
110 East 42nd St., Suite 1300
New York, New York 10017-8532
Tel. (212) 362-7044 • Fax (212) 362-4919
URL: http://www.acsh.org • Email: acsh@acsh.org

Publisher name: American Council on Science and Health
Title: The Next Plague And What Science Will Do To Stop It
Author: Alex Berezow, Ph.D., Hank Campbell, Julianna LeMieux, Ph.D., Steve Schow, Ph.D.
Subject (general): Science and Health
Publication Year: 2018
Binding Type (i.e. perfect (soft) or hardcover): Perfect
ISBN: 978-0-9972530-1-6

Acknowledgements

The American Council on Science and Health (ACSH) appreciates the contribution of the reviewers named below:

Dr. Rudolph Jaeger, DABT, ERT
President
CH Technologies Inc.

Julie Garden-Robinson, Ph.D., R.D., L.R.D., F.A.N.D.
Professor and Food and Nutrition Specialist
North Dakota State University

Jay Lehr
Science Director
Heartland Institute

Olivia Ellis
Doctoral student
University of California at Los Angeles

CSH wishes to thank Cameron English, a freelance science and health writer in Sacramento, CA., for his assistance in fact checking references in this report.

Table of Contents

1 Introduction

Before the study of microbiology, it was not understood what made people sick or how diseases spread.

The leading theory up through about the 1850s, and the understanding of microorganisms, was called "miasma theory." Miasma is the bad-smelling air that originates from decaying material, and it was thought that miasma made people sick. This now disproven theory was the inspiration of the name for the infectious agent that causes over 200 million cases of illness per year – malaria or bad air (*mala*, "bad" and *aria*, "air").

It was not until the advent of the field of microbiology that people started to understand that some diseases are spread by an infection caused by microorganisms. This paradigm shift was proven by the pioneering work of the founders of the field of microbiology, most prominently Antonie van Leeuwenhoek, Louis Pasteur, and Robert Koch.

Although it had been suspected for some time that organisms existed that were too small to be seen by eye (i.e., microscopic), this idea was not proven before the invention of the microscope in the late 17th Century. It was then that the Dutch professional draper, Antonie van Leeuwenhoek, put a drop of water under his handmade microscope and saw the now famous "wee animalcules." In a letter[1] to H. Oldenburg, Oct. 9, 1676 he wrote,

> *"I now saw very distinctly that these were little eels or worms... Lying huddled together and wriggling, just as if you saw with your naked eye a whole tubful of very little eels and water, the eels moving about in swarms; and the whole water seemed to be alive with the multitudinous animalcules. For me this was among all the marvels that I have discovered in nature the most marvellous of all, and I must say that, for my part, no*

more pleasant sight has yet met my eye than this of so many thousands of living creatures in one small drop of water, all huddling and moving, but each creature having its own motion."

From this moment of the construction of a microscope powerful enough to see bacteria, the field of microbiology exploded. Two hundred years later, the work of Louis Pasteur and Robert Koch extended the knowledge of microbes and showed that van Leeuwenhoek's wee animalcules were actually the cause of disease. This theory came to be known as the "germ theory of disease."

Robert Koch defined, through elegant experimentation, the four rules that must be true in order to conclude that a particular microorganism causes a disease, now known as "Koch's postulates":

1. The microorganism must be present in all cases of the disease.

2. The pathogen can be isolated from the diseased host and grown in pure culture.

3. The pathogen from the pure culture must cause the disease when inoculated into a healthy, susceptible laboratory animal.

4. The pathogen must be re-isolated from the new host and shown to be the same as the originally inoculated pathogen.

These postulates are the mainstay of medical microbiology and have been used time and again to prove a causal relationship between a particular type of bacterium and a specific disease. For example, we know the bacterium *Vibrio cholerae* causes cholera and the bacterium *Borrelia burgdorferi* causes Lyme disease.

Now, roughly another 200 years later, we live in an age where the germ theory of disease is widely accepted. For the sake of public health, it is imperative to look toward the future and wonder what the evolution of infectious disease will mean for our civilization.

Specifically, what will be the next plague, and can we prevent it?

2

The evolution of the host-pathogen relationship

Using genetics and molecular clock analysis, modern human infectious diseases can be traced back to anywhere between 10,000 to 2.5 million years ago (the Paleolithic era) and suggest that many of these pathogens co-evolved with our ancestral hominids in Africa.

For example, the bacterium that causes tuberculosis (TB), *Mycobacterium tuberculosis*, appears to have its origins in Africa 70,000 years ago and jumped into hominids around that time.[2] On the other hand, leprosy (caused by *Mycobacterium leprae*) may be a disease of recent origin, given its absence in pre-Columbian Americas. The lack of genetic variability among isolates is another clue to support this hypothesis. However, other data suggest leprosy is as old as 10 million years.[3]

Smallpox might have originated in an African rodent during the Paleolithic period.[4] Zoonotic diseases, those that originated in animals and then spread into humans, might include diseases arising in other primates, including Neanderthals. Typhoid bacteria (*Salmonella typhi*) seem to have entered humans during a time when *Homo sapiens* intermingled with Neanderthals.[5]

Where Do "New" Infectious Diseases Come From?

The numbers are not slowing down, either. Estimates of emerging infectious diseases (those that are new to the human population) between 1940 and 2004 suggest 335 such entities appeared in that period, or about five emerging infectious disease events per year.[6]

Infectious diseases can be categorized in many different ways, one of which is based on their occurrence in the human population. For example, established infectious diseases like TB, malaria, and tapeworms have a constant presence in the population.

Emerging infectious diseases have either not occurred in humans before, have occurred previously but only in small numbers of people that are located in remote places (e.g. Ebola, Lassa, cholera), or have occurred throughout human history, but the cause was previously considered to be something other than an infection (e.g., Lyme disease and gastric ulcers).

Lastly, there are re-emerging infectious diseases, classified as such because they caused health problems in large numbers at one point, then declined, but re-emerged as problematic. Current examples are whooping cough and yellow fever.

New human pathogens must come from somewhere. Unsurprisingly, most come from animals (i.e., are zoonotic). Sixty percent of current human infectious diseases, even those exclusively in humans like measles, mumps, whooping cough, and hepatitis B, are of zoonotic origin, probably coming from wild animals.[7] When microbes that normally colonize animals acquire the attributes that allow them to infect humans, they are able to make the jump into humans, which we refer to as a spillover.

Darwinian evolution optimizes the microbe for adaptation to human hosts. The infectious organisms may become so finely tuned to humans that they become exclusively human pathogens. Classic cases of opportunistic spillover from animals to humans include HIV, monkeypox, and Ebola (from non-human primates); SARS (civet cats and bats); MERS (camels); human Q-fever (dairy goats); and influenza (waterfowl and swine).

All kinds of microorganisms can infect the human body. The four major types are: viruses, bacteria, fungi, and parasites, which include protozoa and helminths.

3

Viruses: The superstars among plague microbes

Viruses are among the most dangerous infectious agents on Earth. They can hijack the cells of any living organism, including bacteria and plants. Though they are efficient molecular machines, paradoxically, viruses are not alive. They do not demonstrate metabolic activity or generate any of the energy molecules of life. They are not self-sustaining, and they do not reproduce outside a live host organism. Viruses are coated with proteins that adhere to their target cells simply by their chemical recognition and affinity properties. While our bodies' cells are encoded by DNA, viruses can be encoded by either DNA or RNA.

Viruses share a common pathogenic strategy: they enter target cells in order to commandeer the host cell's machinery with the sole intention of producing millions of daughter viruses. (Imagine walking into an airplane factory and demanding that everybody stop what they're doing so they can start making buses, instead. That's essentially what a virus does.) These newly synthesized viruses then spread throughout the host's body and eventually back into the environment to start the cycle anew.

The amazing process of viral infection can be demonstrated through the Human Immunodeficiency Virus (HIV), a special type of RNA virus called a retrovirus. HIV is a bag of fat molecules surrounding a watery core. The fat bag is studded with special viral proteins. When the virus sticks to a target cell (usually a T-cell), the virus is triggered to inject its contents into the cell. Like pirates, the viral molecules immediately take charge of the cell and initiate the viral replication process. Amazingly, the

HIV RNA is converted into DNA, which then physically integrates into the cell's chromosomes. Now, the virus is literally a part of the person. Each infected cell produces 1,000 to 10,000 HIV virions in its lifetime.

Using HIV-1 sequences preserved in human biological samples, along with estimates of viral mutation rates, scientists calculated that the jump from chimpanzee to human probably happened during the late 19th or early 20th Century[8], a time of rapid urbanization and colonization in equatorial Africa. By 2015, 36.7 million people were infected by HIV. The total number of HIV-related deaths is 35 million. It was only because of drugs invented in laboratories that the absolute death sentence from an HIV infection has been successfully derailed. Now, HIV is a manageable chronic disease.

Which Viruses Could Be the Next Plague?

HIV is a classic example of a dangerous, emerging, viral disease with its origin in the animal kingdom. The same is true of smallpox, rabies, Spanish flu, polio, and Ebola. These fear-evoking, catastrophic viruses can wipe out entire populations. What else is lurking out there?

The World Health Organization convened a workshop in December 2015 to prioritize a list of highly infectious or otherwise worrisome pathogens.[9] Incidentally, they are all zoonotic viruses.

1. Crimean Congo hemorrhagic fever
2. Filoviral diseases (e.g., Ebola and Marburg)
3. Coronaviruses (e.g., MERS and SARS)
4. Lassa Fever
5. Nipah
6. Rift Valley Fever

7. Chikungunya

8. SFTS ("severe fever with thrombocytopenia syndrome") virus

9. Zika

Any of these viruses could become a big problem. However, a particularly virulent strain of the highly communicable influenza virus is most likely to be our next really big problem.

The damage done by the great 1918 Spanish flu is astonishing. Five hundred million people across the world were infected with the strain, including people in remote Pacific islands and the Arctic. The pandemic is said to have killed 50 to 100 million people (estimated to be three to five percent of the world's population at that time), a number so large that it is difficult to comprehend.[10]

Although less severe, serious influenza pandemics have occurred since that time, including the Asian flu (1957), Hong Kong flu (1968), and the Russian flu (1977).

Though flu is the likeliest virus to be the next plague, the hemorrhagic viruses may trigger the most panic. Hemorrhagic diseases cause massive vascular leakage; i.e., the leaking of blood out of the circulatory system into the body's tissues. Hemorrhagic viruses include Ebola, dengue, Rift Valley fever, Lassa fever, Marburg, Hantavirus, Crimean-Congo virus, and Machupo virus, among others. A particularly frightening one is the rodent-borne Guanarito virus that causes the highly lethal Venezuelan hemorrhagic fever.

Venezuelan hemorrhagic fever (VHF), first identified in 1989, is most prevalent in several rural areas of central Venezuela. The short-tailed cane mouse is the main host of Guanarito virus, which is spread primarily through inhalation of aerosolized droplets of saliva, respiratory secretions, urine, or blood. Although uncommon, people can spread the virus to each other. The disease is fatal in 30% of cases.[11]

It is speculated that VHF emerged following demographic and ecological changes in rural areas that increased the frequency of contact between humans and infected rodents. Because it is so dangerous, some initially speculated that the disease was the result of biological warfare. But, it was completely natural. It is possible that as the Guanarito virus continues adapting within its rodent host, mutations will arise that will allow it to spread more easily among humans. If that were to happen, VHF could become a major threat in the Americas and beyond.

Besides coming from nature, viruses are thought to be ideal biological warfare agents. Even if a bioterrorist incident had a limited impact (in terms of public health), societal disruption would be severe. Just consider how terrified many in American society were following the 2001 anthrax attacks and the 2014 Ebola epidemic. If a smallpox outbreak were to occur – especially if it was purposeful – the resulting panic would disrupt society and the global economy, possibly to the tune of billions of dollars, if history were to repeat itself.[12]

Viruses Have a Unique Penchant for Mutation

Viruses can be very sloppy during replication, which leads to mutation. This is particularly true of viruses that have RNA (rather than DNA) or single-stranded (rather than double-stranded) genomes. The speed with which some viruses mutate can make them challenging for biomedical scientists and public health officials.

Mutation is the mechanism by which natural selection and evolution occur. Through chance, mutations can give viruses new properties, such as the ability to infect different kinds of cells or even species. That's how the influenza virus hops from swine and birds to humans and back again.

Furthermore, mutation is the mechanism by which viruses evolve resistance to antiviral drugs.

Mutations are why some viruses are unlikely to ever be controlled or prevented by a vaccine. HIV, a rapidly mutating virus, serves as a classic example. Even within the same person, there is so much genetic diversity among HIV viruses that researchers refer to it as a "quasispecies." Mutations also explain why we have to get annual flu shots: Flu viruses are constantly mutating, and our immune systems often fail to recognize the new strains that evolve.

Preventing the Next Viral Plague

Viruses can be controlled prophylactically (meaning in a preventative way) with vaccines and antiviral drugs.

Vaccines have a terrific track record at reducing or eliminating many of the most lethal and contagious viruses in the world. Vaccination eradicated smallpox and rinderpest, a disease of cattle. During childhood, kids are vaccinated against a variety of viruses, such as those that cause measles, mumps, rubella, chickenpox, hepatitis, and rotavirus. Cervical cancer can be prevented with a vaccine against human papilloma virus.

Viruses that cannot be controlled with vaccines can be treated with antivirals. HIV, which was at one time a death sentence, is now rather easily controlled with a cocktail of antiretroviral medications. An antiviral drug can cure people infected with hepatitis C.

Therefore, while viruses remain the biggest threat from the microbial world, we have an effective arsenal with which to combat them. But what about bacteria?

4

Bacteria: The coming and going and coming again of bacterial plagues

Unlike viruses, bacteria are living cells that are – with a few exceptions – capable of independent survival in the environment. They come in two broad types, known as Gram-positive and Gram-negative, which refers to the structure of their cell walls.

Bacteria are ubiquitous. Indeed, there are approximately 10^{30} bacteria on Earth.[13] That's more than the number of stars in the universe, which might be 10^{19}, according to some estimates.[14]

Bacteria inhabit every corner of our planet, including every surface we contact, our gardens, our companion animals, our food, and our loved ones. They colonize the surface of our skin, our nose and mouth, and our gut. They live in completely inhospitable places, such as thermal hot springs and Antarctic ice. Some need oxygen, but others are killed by oxygen. Some are pathogenic to humans, while some are helpful; others are benign but can be problematic if given the opportunity. In many ways, bacteria are "frenemies" – can't live with 'em, can't live without 'em.

Some of the worst plagues in human history were due to bacteria: "The Plague," also known as Black Death (*Yersinia pestis*); typhus (*Rickettsia prowazekii*); tuberculosis (*Mycobacterium tuberculosis*); cholera (*Vibrio cholerae*); and leprosy (*Mycobacterium leprae*). Many sexually transmitted diseases are bacterial: Syphilis (*Treponema pallidum*); chlamydia

(*Chlamydia trachomatis*); and gonorrhea (*Neisseria gonorrhoeae*). Bacteria may also serve as potential biological weapons: Anthrax (*Bacillus anthracis*); botulism (*Clostridium botulinum*); and tularemia (*Francisella tularensis*).

Thanks to modern science and medicine, however, humanity has largely triumphed over bacteria.

The Miracle of Antibiotics

During a large typhus fever outbreak in Bolivia in December 1947, Doctor Eugene Payne from the pharmaceutical firm, Parke, Davis & Company, arrived at La Paz General Hospital carrying a small supply of a new drug. Twenty-two patients who were sick with typhus and probably near death were chosen to receive it.[15]

A death certificate, complete except for date, had been made out for Case 10, a man named Gregorio Zalles, and arrangements made for his burial. He had been in a coma for three days when he was given an injection of the new drug. Then the miraculous occurred. He awoke and asked for water. While many others died, Zalles was well in a few days, as were all the others treated with the drug.

What was this lifesaving elixir? The antibiotic chloramphenicol, which was isolated from a culture of the bacterium *Streptomyces venezuelae*.

Unfortunately, this inspiring bit of history about the origins of one of the most important antibiotics ever discovered is not likely to be repeated. If discovered today, rather than in 1947, chloramphenicol would never even make it to the market because its potential side effects include delirium and leukemia. But when people were dying from typhus, huge risks were more palatable.

Consider Salvarsan, an organo-arsenic compound that was released in 1910 to treat syphilis. It was toxic and difficult to administer to patients. However, when death was the alternative, patients opted for Salvarsan.

Because of antibiotics, cases of tuberculosis (the "white plague" and #2 killer in the U.S. in 1900) fell. Sanatoriums, where people with intractable illnesses were held, began closing down by the mid-1950s[16]. The National Leprosarium in Carville, Louisiana, where people with leprosy were treated, was the only holdout and finally closed in 1999.[17]

Bacteria on the Run

Until the 20th Century, bacteria ran wild. The discovery of antibiotics put humans on the offensive and put an end to the microbial reign of terror. Indeed, antibiotics – along with better sanitation and hygiene – were a game-changer for public health. In 1900, according to the CDC[18], the top three causes of death in the United States were due to infectious disease:

1. Pneumonia and influenza
2. Tuberculosis
3. Diarrhea (Gastrointestinal infections)
4. Heart disease
5. Stroke or cerebrovascular disease

Today, CDC figures show that infections don't even make an appearance in the top five causes of death:

1. Heart disease
2. Cancer
3. Chronic airway disease (COPD)
4. Accidents
5. Stroke (Cerebrovascular disease)

Pneumonia, including influenza (a virus), comes in at #8.

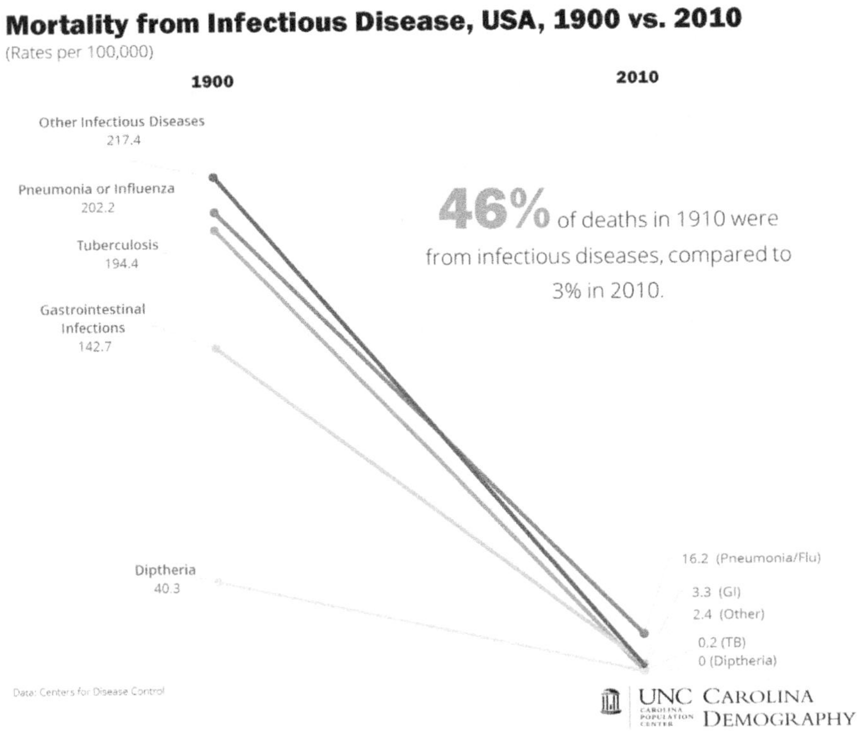

At the end of the 19th century, life expectancy in the U.S. was 47 years on average. By the mid-1970s, that figure rose to 78.8 years, primarily due to the introduction of antibiotics.[19] Watching this dramatic improvement in public health unfold, some medical professionals in the early 1960s described the development of antibiotics as "...one of the most important social revolutions in history," and argued that it had virtually eliminated infectious diseases.[20]

Antibiotic Resistance: The Bugs Strike Back

Unfortunately, these optimistic experts ignored some important facts. In 1940, just 12 years after the introduction of penicillin, scientists identified the first bacterial strain resistant to the drug.[21] Moreover, bacteria began evolving resistance to tetracycline in 1959, and the first case of MRSA (methicillin-resistant *Staphylococcus aureus*) appeared in 1961, just two years after the introduction of methicillin.[21]

Antibiotics play a vital role in routine medical care. Many common procedures are highly dependent on the efficacy of antibiotics. Patients about to undergo surgery are often given antibiotics prophylactically (i.e., to prevent infection). Cancer patients on chemotherapy, autoimmune patients using immunosuppressant drugs, organ transplant recipients, pregnant women undergoing C-sections, and people receiving joint replacements all rely on antibiotics.

Without effective antibiotics, the risk of infection from these and other medical procedures is simply too high. It's not an exaggeration to conclude that a world without effective antibiotics would send modern medicine back to the Victorian Age.

The biggest bacterial threats, according to the CDC, are: *Clostridium difficile* (which causes diarrhea), carbapenem-resistant *Enterobacteriaceae* (which cause blood infections), and *Neisseria gonorrhoeae* (the cause of gonorrhea).

The World Health Organization has its own priority list, as well:

Priority 1: CRITICAL

- *Acinetobacter baumannii* (carbapenem-resistant)
- *Pseudomonas aeruginosa* (carbapenem-resistant)
- *Enterobacteriaceae* (carbapenem-resistant, 3rd generation cephalosporin-resistant)

Priority 2: HIGH

- *Enterococcus faecium* (vancomycin-resistant)
- *Staphylococcus aureus* (methicillin-resistant, vancomycin-intermediate and resistant)
- *Helicobacter pylori* (clarithromycin-resistant)
- *Campylobacter* spp. (fluoroquinolone-resistant)
- *Salmonellae* (fluoroquinolone-resistant)
- *Neisseria gonorrhoeae* (3rd generation cephalosporin-resistant, fluoroquinolone-resistant)

Priority 3: MEDIUM

- *Streptococcus pneumoniae* (penicillin-non-susceptible)
- *Haemophilus influenzae* (ampicillin-resistant)
- *Shigella* spp. (fluoroquinolone-resistant)

The threat of antibiotic resistance is especially critical in the case of Gram-negative bacterial pathogens, such as *Pseudomonas aeruginosa, Acinetobacter baumannii*, and *Klebsiella pneumoniae*. Infections caused by "multidrug-resistant" (MDR) Gram-negative bacteria result in higher mortality rates, longer hospital stays, and higher treatment costs compared to antibiotic-susceptible infections. Patients with MDR Gram-negative bacterial infections, caused by organisms including *Pseudomonas, Serratia, Acinetobacter* and *Enterobacter*, have a mortality rate of 30-70%.[22]

Gram-positive bacteria are also worrisome. MRSA, for instance, was widespread in Europe by the end of the 1970s and in the US by the end of the 1980s. Healthy people can carry MRSA in their noses, which serves as a source of hospital-acquired infections. In 2005, there were 368,600 hospital stays with MRSA infections in the U.S., resulting in 18,650 deaths and costing $5 billion.[23] Fortunately, the number of MRSA infections declined nearly 30% between 2005 and 2011, and killed 9,000 fewer hospital patients in 2011, according to the most recent data available from the CDC.[24]

One of the biggest threats to public health worldwide is the explosion of drug-resistant tuberculosis. More than 50 years after the introduction of the first of many successful anti-TB drugs, the disease infects 10.4 million people annually, and kills 1.7 million of them. More than 95% of these deaths occur in developing nations like India, Indonesia, China and several others.[25]

Currently, antibiotic-resistant bacterial infections are most problematic for hospitalized patients, the elderly, or individuals with some sort of immune suppression or deficiency. The CDC says that 2 million people acquire antibiotic-resistant infections every year in the U.S. – many of which occur in hospitals – and roughly 23,000 die from them.[21] (As it turns out, if you're really sick, the hospital is the last place you want to be.) Furthermore, these infections incur an estimated $20 billion in direct healthcare costs and $35 billion in lost productivity, both of which are expected to rise rapidly over the next several years.[26]

What can we do?

Preventing the Next Bacterial Plague

Unfortunately, the pharmaceutical industry began to abandon antibiotic research as the 1980s came to a close. Between 1983 and 1987, 16 new antibiotics were approved in the U.S. By comparison, only two new drugs reached the market between 2008 and 2011.[30] The Food and Drug Administration established new clinical trial requirements in the 1990s[27] which, among other things, roughly doubled the number of enrollees needed for an antibiotic clinical trial. This made it prohibitively expensive and sometimes logistically impossible to successfully register a new antibacterial drug.

A recent law enacted in December 2016, the 21st Century Cures Act, directs the FDA to facilitate the introduction of new antibiotics. The trouble, however, is that it could take decades for new drugs to show up in pharmacies. Additionally, the economics for inventing new antibiotics are not expected to change. The estimated cost to develop a new anti-TB drug, for instance, is as high as $300 million.[28] Furthermore, governments severely restrict the use of antibiotics to limit drug resistance, which makes development much less profitable.[29] In other words, there is no strong financial incentive to encourage antibiotic discovery.

To prevent the next bacterial plague, therefore, we need innovative policies. But that's not all.

For hospitals, infection control is key. Simple things – like making sure that doctors and nurses wash their hands – can go a long way. Technological advances, which incorporate antibacterial surfaces into medical devices such as catheters, can also help stop the spread of bacteria.

From a societal standpoint, judicious use of antibiotics is mandatory. Antibiotics should never be prescribed for the sniffles; most likely, that's a viral infection. Scientists, of course, are working tirelessly to fight antibiotic resistance. They are seeking not only new and unusual antibiotics, but new and unusual bacterial targets to hit with them.

Though bacteria are mostly managed, they are staging a comeback. The longer-term paradox is determining how to fight bacteria without worsening antibiotic resistance. While antibiotic resistance should not induce panic, it is still a serious threat that will require out-of-the-box thinking.

5

Fungi: If you run afoul of fungi, it's a plague to you

Like bacteria, fungi are living cells that are capable of surviving independently in the environment. Fungi, and the spores they produce, are ubiquitous. Fungi also come in a wide variety of shapes and sizes, from the microscopic (i.e., not visible to the human eye) to the macroscopic (i.e., visible to the human eye). Some of our favorite foods, like mushrooms, are types of fungi. Other foods, like bread, cheese, and beer, rely on fungi.

As a general rule, fungi are not a threat to healthy, immune competent individuals. Therefore, they are not considered a pandemic threat. However, people who are immunocompromised are susceptible to opportunistic fungal infections. (As its name suggests, fungi cause infections when given the right opportunity.)

HIV and drug-induced immunosuppression (for treating transplant, cancer, and autoimmune disease patients) highlights the opportunistic nature of fungi. In the early days of the AIDS crisis, many patients were dying from rare secondary infections, such as from the fungus *Pneumocystis carinii* (now known as *Pneumocystis jirovecii*). Because HIV targets T-cells, a critical component of the immune response, patients succumbed to diseases, such as those caused by fungi, that healthy people never get.

The Fungus Among Us

By far, the biggest global threat from fungus is aflatoxin. When insects munch on crops, they create wounds that can become infected with a type of fungus called *Aspergillus*. *Aspergillus* produces a toxin, called aflatoxin, that can cause cancer. As a result, the food supply is closely monitored to keep the concentration of aflatoxin as low as possible.

Agricultural fields are vulnerable to fungal attack because they are large and genetically homogeneous. It is possible, therefore, that a terrorist could spray crops with *Aspergillus* or another toxin-producing fungus. While aflatoxin can be lethal in a large enough dose, the risk of human casualties would be quite low. The actual threat from crop contamination, therefore, is the disruption that would result from destroying part of the food supply.

In certain communities, fungi can be problematic. People who live in the arid western U.S. or Mexico come into frequent contact with the fungus that causes San Joaquin Valley Fever (or just "Valley Fever"). Chances are that if you ever have lived in this area, you have been exposed to this fungus. Most likely, the exposure either went unnoticed (because your immune system worked well) or manifested as flu-like symptoms which then went away.

About 12,000 cases of Valley Fever are reported annually.[31] In Arizona and California alone, approximately 2,500 people are hospitalized each year, with costs exceeding $50 million.[32] However, hospitalization data from recent years is limited.[33] For some patients, there is no practical solution, as they are unresponsive to the available antifungal drugs or cannot tolerate their side effects. Occasionally, and particularly in immunocompromised individuals, the immune system loses control of the fungus, and the infection wreaks havoc in the patient. This is rare, however, since the infection kills fewer than 200 people a year.[31]

Our four-legged friends are also vulnerable to San Joaquin Valley Fever. The fungus infects 6-10% of dogs living in Pima, Pinal, and Maricopa counties in Arizona every year.[34]

Other fungal exposures occur as the result of accidental contamination. In September 2012, an outbreak of fungal meningitis was reported in the United States, which the CDC traced to contamination in three lots of a medication called methylprednisolone used for epidural steroid injections. The drug had been administered to about 14,000 patients. Because of the unusual nature of the infection, it took clinicians and public health officials a few months to determine the cause of the outbreak. As of October 29, 2015, 64 people had died and 753 were being treated for persistent fungal infections.[35]

In November 2012, some patients recovering from meningitis were reported to be experiencing secondary infections at the injection site. A black mold called *Exserohilum rostratum* was found in 150 of the cases and *Aspergillus fumigatus* was found in one case.[35] Cases of meningitis caused by *Aspergillus* are rare, but cases caused by black mold are even more so. Both of these fungi are common in the environment and rarely cause problems for humans, unless a person is immunocompromised or injects the fungi directly into the body.

Preventing the Next Fungal Plague

Because fungi pose little pandemic threat, there isn't much to be done to prevent it. Antifungal drugs, though unpleasant, already exist.

Anti-fungal efforts ought to be focused on agriculture and maintaining a safe food supply. The agrichemical industry is the main source of novel fungicides, which are used to treat plant fungal pathogens. Researchers also are developing crops that are less vulnerable to fungal infection. Some GMOs (such as those containing Bt toxin) kill insects and reduce the potential for fungal infection, while others "turn off" the fungal genes required to produce aflatoxin.

Overall, there is little reason to worry about a renegade fungus. But what about parasites?

6

Parasites: The massive ongoing plagues none of us care about

Parasites are organisms that live primarily by harming other organisms. Like fungi, parasites come in all shapes and sizes, from single-celled creatures to gigantic worms.

Because the world's worst parasites are mostly a problem for poor countries, they have largely been ignored. Often called "neglected tropical diseases," parasitic infections have recently received more attention due to the efforts of philanthropic organizations such as the Bill & Melinda Gates Foundation.

The most well-known and devastating parasite is malaria, a single-celled organism. The numbers are truly staggering. According to the World Health Organization, half the world's population is vulnerable to malarial infection, more than 200 million people are infected annually, and more than 400,000 of them die.[36] Humans are the main host reservoir for malaria, which is transmitted by the female *Anopheles* mosquitoes.

The world's second biggest parasitic killer is amoebic dysentery (bloody diarrhea), caused by the single-celled microbe *Entamoeba histolytica*. This organism sickens 50 million people and kills 40,000-100,000 annually.[37] Yet another single-celled parasite, *Leishmania*, is transmitted by a sandfly bite and is responsible for up to 1 million infections and 30,000 deaths every year.[38]

The nastiest parasites – from a visual perspective – are worms. Guinea worms, which cause dracunculiasis, emerge from an infected person's skin, often a foot. Elephantiasis, in which a person's legs can swell to a grotesque size, is caused by a worm that blocks lymph vessels.

Not all parasites live overseas. The single-celled parasite *Cryptosporidium* causes diarrhea. In 1993, the organism made an appearance in the water supply of Milwaukee, Wisconsin, sickening some 400,000 people and killing 69.[39]

The CDC believes that *Toxoplasma gondii*, a single-celled parasite, chronically infects 30 million Americans. Toxoplasmosis can be acquired through eating undercooked meat or handling the feces of a pet cat. Because our immune system keeps *Toxoplasma* in check, it's not a problem for those who carry the parasite. However, a pregnant mother who is infected may see serious consequences in her baby.

Preventing the Next Parasitic Plague

Parasites are not much of a problem in the developed world, partially because of an advanced health infrastructure and partially because the climate is not suitable to some of them. The parasites that do exist in the U.S., such as *Toxoplasma*, are nowhere nearly as devastating as malaria. However, if you are one of those unlucky people who acquires a serious parasitic infection, if the parasite doesn't kill you, the treatment may make you wish it had.

Part of the reason for this is that there hasn't been much in the way of anti-parasitic drug development in recent decades.[40] Many of these drugs were developed in the first half of the 20th Century to prevent troops from acquiring exotic illnesses, but by the 1980s, most pharmaceutical companies had shuttered their anti-parasitic drug research. Therefore, we are frozen in time with respect to anti-parasitic treatment advances. This is a potential problem, since malaria is already resistant to multiple drugs.

Unfortunately, pharmaceutical companies show little interest pursuing this line of research because there is little money to be made curing poor people. They could never recoup their investment, so they decide to pursue treating more lucrative diseases with their limited pool of R&D funds.

The bright side is that vector control is effective. Killing mosquitoes prevents them from spreading malaria and other diseases. This can be achieved through application of DDT, which the World Health Organization still recommends even for indoor use[41], or through "high-tech" solutions such as releasing genetically modified mosquitoes into the environment that produce nonviable offspring. Hypothetically, this latter approach could allow scientists to target specific disease-carrying mosquito species for extinction.

Now and for the foreseeable future, it is primarily citizens of the developing world, immunosuppressed patients, and global travelers who are most at jeopardy from parasitic infections. For these souls, the "next parasitic plague" is less worrisome than the myriad of current parasitic plagues.

7

Parasitic worms: the greatest plague of all

Parasitic worms, also called helminths, include organisms like roundworms (nematodes) and flatworms (such as flukes and tapeworms). Worms will not become the *next* devastating global plague because they *currently* own that title.

Shockingly, it is likely that in excess of 1.5 billion people are infected with parasitic worms, many of which are transmitted through contaminated soil, water, and food.[42] An exact death toll is difficult to come by, but parasitic worms may kill 500,000 people annually, and that figure is probably too conservative.[43]

Parasitic infections can be visually striking. For instance, guinea worms (the cause of dracunculiaisis) emerge from sores on a person's skin. Major global parasitic infections include:

- Lymphatic filariasis (elephantiasis), according to the CDC, infects over 120 million people a year and results in the disfigurement or incapacitation of 40 million.[44] The disease can cause grotesquely swollen limbs, which are not only painful but disfiguring and may lead to social stigmatization.

- Loiasis (African eye worm) infection results in itchy, non-painful swellings commonly found near joints. Worms may crawl across the surface of the eye or under the skin.

- Onchocerciasis (river blindness) brings with it eye lesions that itch. After years, the lesions may lead to blindness and skin diseases (which result in "leopard" skin or "lizard" skin).

- Schistosomiasis, caused by blood flukes, results in abdominal pain, diarrhea, bloody stool, and an enlarged liver and spleen. There may be blood in the urine, and bladder cancer is a possible complication.

The pharmaceutical industry abandoned the anti-helminthic drug business long ago. Just like with other parasites discussed in the previous chapter, there just isn't much money to be made curing poor people in faraway lands. Fortunately, the need to treat helminthic diseases in livestock and companion animals has generated interest in producing drugs that can also be used in humans.[45]

The drugs that do exist are often harsh. For example, diethylcarbamazine (DEC) is the drug of choice for treating lymphatic filariasis. Common side effects are itching, facial swelling, headaches, and fatigue. Vision loss is also possible.

Thankfully, the developed world has little to fear from parasitic worms. There are only two ways that you are likely to encounter one: Bad luck or adventurous travel.

Anisakidae, sometimes called herring worms, are roundworms living in various seafood species. During the worm's lifecycle, it passes through small crustaceans, then into fish or squid, and finally landing in the gut of a sea mammal. Thus, undercooked seafood or sushi can expose people to this helminthic parasite. Sometimes, the worm is regurgitated, while others have to undergo surgery to remove it.

Angiostrongylus cantonensis is a roundworm that commonly resides in the pulmonary arteries of rats, giving it the nickname rat lungworm. Snails are the primary intermediate hosts, where larvae develop until they are infective. Humans may become infected through ingestion of larvae in raw or undercooked snails or from contaminated water or vegetables. This worm originated in Asia but is now endemic in Hawaii, where it killed at least one person.

Off-the-beaten-path adventurers who like to travel to impoverished countries, particularly those in the tropics, should pay close attention to strange or bewildering symptoms. Such an excursionist might have picked up a new friend. Unfortunately, once the intrepid traveler comes home, a general practitioner may have a hard time diagnosing the medical condition, given the rarity of helminth infections in advanced countries.

Preventing the Next Parasitic Worm Plague

There's little point in worrying about the next parasitic plague because the current one is so incredibly devastating. Fortunately for those of us in the developed world, parasitic worms are not – and will not be – a daily fact of life.

But what about mad cow disease?

8

Infectious proteins: Prions and transmissible spongiform encephalopathies

Like viruses, prions aren't alive. Unlike viruses, they don't even contain DNA or RNA. Instead, a prion is just a misfolded molecule, specifically a type of protein. As such, they are the simplest infectious agents of humans.

Your brain already has prions in it – although, if the proteins are folded normally, they aren't called prions. These proteins, whose function is not entirely understood, are found largely in the central nervous system. The proteins can be converted into prions when they come into physical contact with another prion. In other words, a misfolded prion protein triggers normal proteins to misfold. This begins something of a chain reaction that results in the prions clumping together and leading to neurodegeneration.

Where do these preexisting misfolded prions come from? They can either occur spontaneously in the brain – either from bad luck (a misfolding event) or bad genetics (which predispose a person to misfolding events) – or they can be introduced, such as by consuming tainted food. In humans, the condition is known as Creutzfeldt-Jakob disease.

The outbreak of "mad cow disease" (more properly known as bovine spongiform encephalopathy) in the United Kingdom occurred after people consumed beef that contained prions. Millions of cattle had to be euthanized. While thousands of people in Britain and elsewhere were exposed, only 231 people became infected.[46] However, the fatality rate is 100%.

The reason so many cattle were potentially affected was because they consumed contaminated animal feed. At the time, animal feed was allowed to contain whatever animal parts humans didn't want, such as nervous system tissue. Cows that died of bovine spongiform encephalopathy, therefore, were chopped up and fed to millions of other cows. Since 1997, the FDA has banned this practice.[46]

Prions are not the same as toxins. Toxins are not transmissible, while prions can be transmitted from animals to humans or from humans to humans. Ritual cannibalism among the Fore tribe in Papua New Guinea resulted in the development of kuru, the human equivalent of mad cow disease, after people ate the brains of deceased relatives, at least one of whom probably died from Creutzfeldt-Jakob disease.

Having said that, prion disease is not "contagious" in the traditional sense. There is no evidence to suggest it can be spread from person to person, even by close contact. Once a person or animal has developed prion disease, central nervous system tissues (e.g., the brain, spinal cord, and eyes) are thought to be extremely infectious. However, that is only relevant for those handling infected tissue directly, which does not include caregivers looking after a person with the disease. People with any form of prion disease are requested not to be blood or organ donors and to inform their doctor and dentist prior to any invasive medical or dental procedures.

There's a good reason for that. Prions are not destroyed by conventional sterilization procedures. It is possible to transmit the disease if surgical instruments, perhaps used in neurosurgery, are not properly cleaned.

Humans and cows aren't the only animals affected by this disorder. Others include chronic wasting disease in deer and elk, scrapie in sheep or goats, feline spongiform encephalopathy in cats, and exotic ungulate encephalopathy in some zoo animals.

One of the characteristics that makes prion disease so insidious in humans is its decades-long incubation period.[47] Indeed, scientists have reason to believe that more human cases of mad cow disease from the UK outbreak will occur, even though the exposure happened many years ago. Upon autopsy, those who succumb to prion disease will display spongiform ("Swiss cheese") brain lesions.

Preventing the Next Prion Plague

Given the rarity of these diseases and their unusual mode of transmission, should prion disease be a concern? Yes.

Exposure to infected American bush meat, for example elk or deer with chronic wasting disease, could lead to an outbreak of human prion disease, assuming that elk or deer prions can cause human brain proteins to misfold. Furthermore, the meat industry must remain vigilant for any sign of prion disease in herds. Unfortunately, prions are robust and can persist in the soil for years, which no doubt explains outbreaks of scrapie among sheep that feed in fields where prior scrapie outbreaks occurred. Sheep, as it turns out, shed prions in their feces.

Overall, the likelihood of a prion plague is quite small. And avoiding prion is fairly straightforward: Don't eat another human's brain. Don't eat elk or deer brains. Don't let cows eat cow brains. Bad luck, however, is largely out of a person's control.

Epilogue
Perspectives on the future

In the 21st Century, the developed world has largely conquered infectious diseases. But they have most certainly not disappeared. Because these microbes are constantly evolving with humans, we must remain ever vigilant for the next plague.

Viruses, perhaps, pose the greatest public health threat, and it would not be surprising to see a devastating global influenza pandemic in our lifetime. Deadly bacteria are making a comeback via antibiotic resistance. Ebola has reminded us to keep an eye on zoonotic viruses that emerge into the human population. And synthetic biology teaches us that microbes – even long eradicated ones – can be resurrected and modified into biological weapons in the laboratory.

These are frightening scenarios. However, there are many reasons to be hopeful that science will prevent the next plague.

The first is awareness. Science is no longer naïve. We fully understand the consequences of profligate antibiotic use and that a local outbreak can quickly spiral out of control. In a highly connected world, an Ebola outbreak in a remote African village can become a global pandemic. Experts at the World Health Organization and Centers for Disease Control and Prevention conduct global surveillance in order to prevent this.

The second is vaccination. Simply put, vaccines work. Smallpox, which is thought to have killed possibly 500 million people in the 20th Century alone, has been eradicated from the Earth. Polio and measles may be next. Rinderpest, a viral disease of cattle, has also been eradicated. While it is not possible to eradicate all infectious diseases, vaccines can mitigate the damage they cause. Researchers are working on a "universal influenza vaccine" that, in theory, should protect humans from any type of influenza virus. In late 2016, scientists announced the creation of a highly effective Ebola vaccine.

The third is wealth. Poverty is the real underlying cause of many infectious diseases. As poorer parts of the world become richer, their public health infrastructure will improve. More than one billion people lack access to electricity, which means that they also do not have access to reliable healthcare. Malnutrition leaves people more susceptible to infectious disease. These problems will disappear as economic and technological advancement continues its march across the globe.

The fourth is philanthropy. While it is true that pharmaceutical companies have little incentive to produce drugs that cure poor people (who can't afford them), philanthropists are not held back by such financial constraints. The Bill and Melinda Gates Foundation, for instance, has taken on "neglected tropical diseases," a group consisting mostly of parasitic infections. Collaborating with government and industry, philanthropists will make a sizeable impact in stopping the next plague.

The fifth is biotechnology. Our command over biology has increased exponentially in recent decades. AIDS, a disease that once seemed incurable, is now manageable and those suffering from it live somewhat normal lives. Antibiotic resistance can be tackled with advances in other fields, such as nanotechnology. Specialized nanosurfaces can serve as antibacterial coatings on medical devices. Synthetic biology, in which entirely new genetic and metabolic pathways can be engineered from scratch, appears to be limited only by our imagination.

While none of these factors serve as foolproof insurance against another plague, they give humanity an immense advantage over infectious disease. Because microbes will never stop evolving, it is our mission to always remain one step ahead.

Endnotes

1 A.R. Hall. The Leeuwenhoek Lecture, 1988. Notes Rec. R. Soc. Lond. 43, 249-273 (1989) http://rsnr.royalsocietypublishing.org/content/roynotesrec/43/2/249.full.pdf

2 I.C.B. Comas, et al. Out-of-Africa migration and Neolithic Coexpansion of Mycobacterium Tuberculosis with Modern Humans. Nature Genetics 45(10): 1176–1182 (2013) doi: 10.1038/ng.2744

3 Xiang Y. Han , Francisco J. Silva PLoS Negl Trop Dis. 8(2): e2544 (2014) doi: 10.1371/journal.pntd.0002544

4 Gabriel Trueba. The Origin of Human Infectious Diseases. (2014) https://www.researchgate.net/publication/268979334_the_origin_of_human_infectious_diseases

5 A. J. Sams et al., Adaptively Introgressed Neandertal Haplotype at the OAS Locus Functionally Impacts Innate Immune Responses in Humans. Genome Biology 17(246) (2016) https://doi.org/10.1186/s13059-016-1098-6

6 C.D. Butler. Infectious Disease Emergence and Global Change: Thinking Systemically in a Shrinking World. Infectious Diseases of Poverty 1(5) (2012) https://doi.org/10.1186/2049-9957-1-5

7 D Grace, et al. Mapping of poverty and likely zoonoses hotspots. Zoonoses Project 4. Report to the UK Department for International Development (2012) https://cgspace.cgiar.org/bitstream/handle/10568/21161/ZooMap_July2012_final.pdf?sequence=4&isAllowed=y

8 M.Worobey, et al. Direct Evidence of Extensive Diversity of HIV-1 in Kinshasa by 1960. Nature 455(7213): 661–664. (2008) doi:10.1038/nature07390

9 WHO publishes list of top emerging diseases likely to cause major epidemics. World Health Organization (2015) Accessed: 12-Dec-17. http://www.who.int/medicines/ebola-treatment/WHO-list-of-top-emerging-diseases/en/.

10 JK Taubenberger, DM Morens. 1918 Influenza: the Mother of All Pandemics. Emerg Infect Dis 12(1):15-22. (2006) doi: 10.3201/eid1201.050979

11 N. D. Manzione, et al. Venezuelan Hemorrhagic Fever: Clinical and Epidemiological Studies of 165 Cases. Clinical Infectious Diseases 26(2): 308–313 (1998) https://www.ncbi.nlm.nih.gov/pubmed/9502447

12 K. Kupferschmidt. How Canadian researchers reconstituted an extinct poxvirus for $100,000 using mail-order DNA. Science (2017) Accessed: 12-Dec-17 doi:10.1126/science.aan7069

13 P.D. Schloss, J. Handelsman. Status of the Microbial Census. Microbiology and Molecular Biology Reviews68(4), 686–691 (2004) https://www.ncbi.nlm.nih.gov/pmc/articles/PMC539005/

14 Tia Ghose. How Many Stars Are in the Universe? LiveScience (2016) Accessed: 12-Dec-17, www.livescience.com/56634-how-many-stars-are-in-the-universe.html.

15 Eugene Payne, et al. Treatment of epidemic typhus with chloromycetin. Transactions of the Royal Society of Tropical Medicine and Hygiene 42(2): 163–170 (1948) http://www.sciencedirect.com/science/article/pii/0035920348900212.

16 Thomas M. Daniel. The History of Tuberculosis. Respiratory Medicine 100(11):1862–1870 (2006) doi: http://dx.doi.org/10.1016/j.rmed.2006.08.006

17 History of the National Hansen's Disease (Leprosy) Program. Health Resources & Services Administration. Accessed: 13-Dec-17 https://www.hrsa.gov/hansens-disease/history.html

18 Rebecca Tippett. Mortality and Cause of Death, 1900 v. 2010. UNC Carolina Demography(2014) Accessed 13-Dec-17 http://demography.cpc.unc.edu/2014/06/16/mortality-and-cause-of-death-1900-v-2010/

19 W.A. Adedeji. The Treasure Called Antibiotics. Ann Ib Postgrad Medv.14(2): 56-57 (2016). https://www.ncbi.nlm.nih.gov/pmc/articles/PMC5354621/

20 Brad Spellberg, David N. Gilbert. The Future of Antibiotics and Resistance: A Tribute to a Career of Leadership by John Bartlett. Clin Infect Dis. 15(59): S71–S75 (2014) doi: 10.1093/cid/ciu392

21 About Antimicrobial Resistance. CDC (2015) Accessed: 14-Dec-17 https://www.cdc.gov/drugresistance/about.html

22 Kishor C. Mehta, et al. Burden of Antibiotic Resistance in Common Infectious Diseases: Role of Antibiotic Combination Therapy. J Clin Diagn Res. 8(6): ME05–ME08 (2014) doi: 10.7860/JCDR/2014/8778.4489

23 Anne Elixhauser, Claudia Steiner. Infections with Methicillin-Resistant Staphylococcus Aureus (MRSA) in U.S. Hospitals, 1993–2005. Healthcare Cost and Utilization Project (HCUP) Statistical Briefs (2007) https://www.ncbi.nlm.nih.gov/books/NBK61977/

24 R. Dantes, et al. National burden of invasive methicillin-resistant Staphylococcus aureus infections, United States, 2011. JAMA Intern Med 173(21):1970-8 (2013) doi: 10.1001/jamainternmed.2013.10423.

25 Tuberculosis Fact Sheet. World Health Organization (2017) Accessed: 14-Dec-17. http://www.who.int/mediacentre/factsheets/fs104/en/

26 Antibiotic Resistance Threats in the United States, 2013. CDC (2013) Accessed: 14-Dec-17. https://www.cdc.gov/drugresistance/threat-report-2013/index.html https://www.cdc.gov/media/releases/2013/p0916-untreatable.html

27 Suzanne White Junod. FDA and Clinical Drug Trials: A Short History. FDA (2016) Accessed: 14-Dec-17

28 DT Jamison, et al., editors. Disease Control Priorities in Developing Countries. 2nd edition. Oxford University Press. (2006) https://www.ncbi.nlm.nih.gov/books/NBK11718/

29 Elizabeth Sukkar. Why are there so few antibiotics in the research and development pipeline? The Pharmaceutical Journal (2013) Accessed: 14-Dec-17. http://www.pharmaceutical-journal.com/news-and-analysis/features/why-are-there-so-few-antibiotics-in-the-research-and-development-pipeline/11130209.article

30 Ibid.

31 Valley Fever (Coccidioidomycosis) Statistics. CDC (2017) Accessed: 15-Dec-17 https://www.cdc.gov/fungal/diseases/coccidioidomycosis/statistics.html

32 Gail Sodemeyer, et al. Coccidioidomycosis-associated Hospitalizations, California, USA, 2000–2011.Emerg Infect Dis. 19(10): 1590–1597 (2013) doi: 10.3201/eid1910.130427 Valley Fever 2014 Annual Report. Arizona Department of Health (2014) Accessed: 15-Dec-17 http://vfce.arizona.edu/sites/vfce/files/valley_fever_2014_annual_report.pdf

33 Jennifer Brown, et al. Coccidioidomycosis: epidemiology. Clin Epidemiol. 5: 185–197 (2013) doi: 10.2147/CLEP.S34434

34 Valley Fever in Dogs. University of Arizona Health Sciences. Accessed: 15-Dec-17 http://vfce.arizona.edu/valley-fever-dogs/how-dogs-get-valley-fever

35 Multistate Outbreak of Fungal Meningitis and Other Infections – Case Count. CDC. (2015) Accessed: 15-Dec-17 https://www.cdc.gov/hai/outbreaks/meningitis-map-large.html

36 Malaria Factsheet. World Health Organization. (2017) Accessed: 15-Dec-17 http://www.who.int/mediacentre/factsheets/fs094/en/

37 NC Chavez-Tapia, et al. Image-guided percutaneous procedure plus metronidazole versus metronidazole alone for uncomplicated amoebic liver abscess. Cochrane Database of Systematic Reviews 1. Art. No. (2009) CD004886. doi: 10.1002/14651858.CD004886.pub2

38 Leishmaniasis Factsheet. World Health Organization (2017) Accessed: 15-Dec-17 http://www.who.int/mediacentre/factsheets/fs375/en/

39 Stephen Gradus. Milwaukee, 1993: The Largest Documented Waterborne Disease Outbreak in US History (2014) Accessed: 15-Dec-17 http://www.waterandhealth.org/milwaukee-1993-largest-documented-waterborne-disease-outbreak-history/

40 Parasites - Neglected Parasitic Infections (NPIs). CDC (2017) Accessed: 15-Dec-17 https://www.cdc.gov/parasites/npi/

41 The use of DDT in malaria vector control. WHO position statement. World Health Organization (2011) Accessed: 15-Dec-17 http://www.who.int/malaria/publications/atoz/who_htm_gmp_2011/en/

42 Soil-transmitted helminth infections. World Health Organization (2017) Accessed: 18-Dec-17 http://www.who.int/mediacentre/factsheets/fs366/en/

43 Peter Hotez, Jennifer Herricks. One Million Deaths by Parasites. Plos: Speaking of Medicine (2015) Accessed: 18-Dec-17 http://blogs.plos.org/speakingofmedicine/2015/01/16/one-million-deaths-parasites/

44 Emily Carlson. Taking the 'Bite' Out of Vector-Borne Diseases. NIH (2013) Accessed 18-Dec-17 https://publications.nigms.nih.gov/insidelifescience/vector-borne-diseases.html

45 PJ Brindley, et al. Helminth Genomics: The Implications for Human Health. PLoS Negl Trop Dis 3(10): e538. (2009) https://doi.org/10.1371/journal.pntd.0000538

46 Variant Creutzfeldt-Jakob Disease (vCJD). CDC (2015) Accessed: 18-Dec-17 https://www.cdc.gov/prions/vcjd/risk-travelers.html

47 Erik Z Krause. Variant Creutzfeldt-Jakob Disease and Bovine Spongiform Encephalopathy Clinical Presentation. Medscape (2016) Accessed: 18-Dec-17 https://emedicine.medscape.com/article/1169688-clinical

ACSH

EXECUTIVE TEAM

BOARD OF TRUSTEES

CHAIRMAN

VICE CHAIRMAN

PRESIDENT

MEMBERS

FOUNDERS CIRCLE

BOARD OF SCIENTIFIC AND POLICY ADVISORS

BOARD OF SCIENTIFIC AND POLICY ADVISORS (CONTINUED)

BOARD OF SCIENTIFIC AND POLICY ADVISORS (CONTINUED)

BOARD OF SCIENTIFIC AND POLICY ADVISORS (CONTINUED)

he opinions expressed in ACSH publications do not necessarily represent the views of all members of the ACSH Board of Trustees, Founders Circle and Board of Scientific and Policy Advisors, who all serve without compensation.

Made in the USA
Monee, IL
04 November 2026

67376415R00036